'Zeus Genset' (generator and engine)

Author:

Dr. *Antonio Silvestro*, self-employed heavenly based.
Department of *'Mathematic, Physics and Natural Sciences'*, Ba *'Biological Sciences'* University Federico II of Naples, Naples (NA), 80100, Italy.
Department of *'Agriculture'*, MSc *'Plant, Food Science and Environmental Biotechnology'*, University Federico II of Naples, Portici (NA), 80055, Italy.

Abstract:

The introduced document is act for sharing *'Zeus'*, the innovative prototype of Genset (stepper motor + generator) theoretical calculi and sketches suitable for 3D-Printing modelling, based on the *Singularity* that gave origin to the Bubble Universe.

Keywords: engineering, motor, engine, generator, astronomy, mechatronics, chemistry, biology.

Correspondence for Copyright © permissions requirement to:

Dr. Antonio Silvestro born Friday 15[t h] May 1992 at 20:00 under the Taurus sign ascendant Skorpio according the Greek specular to Monkey 猴 hóu rising dog (*Canis lupis*) Canis Major狗 gǒu in Chinese and chestnut (*Castanea sativa*) in Druid, flower (*xochiti*) in Aztec astrology, North knot Capricorn goat (*Capra hircus*) 羊 yang and South knot Cancer rat (*Ractus norvegiensis*) 鼠 shǔ, resident in n°100 Nazario Sauro St., 80026, Casoria (NA) (Italia), number phone: +39 3382634244, emails: dr.antoniosilvestro@gmail.com, tonysilverxxx@gmail.com and antonio.silvestro5@studenti.unina.it.

Index

Architecture and Design: material and methods

In an *isolated system*, neither matter nor energy can be exchanged with the surrounding environment incapable to interact neither constructively nor destructively with this last; hence, do energy and matter see themselves constantly conserved. Despite the aforementioned, *gravity*, albeit in insignificant amount, leave the only force that can penetrate these non-reactive arrangements. Is it the omnipresent Goddess of the physics? Object in the Bubble Universe as you and I are under its action, that leave us grounded. Just inhaling the vital energy, we can feel ourselves a bit freer from her. Gravity, would the constant degree of freedom, onto all the other parameters depend, but certainly, for pragmatic calculi of a routinely ingenious consider it could be superfluous. Bright, by its own nature. Invisible, when lonely. Empty when in love. The same, filling the feelings among the intriguing Greek deities living romantic relationships, sometimes in need of deeper isolation as the hermetic *Zeus* Genset cylinder designed with variable size could be used also for other application such as vacuum pump-compressor, hyper-capacitor, battery and even Green Houses.

A dipole in the core of *Zeus* **rotor** of a motor have to spin $\alpha = 51°$ for avoiding magnetic turbulences damaging *Hera* **stator** and the overall motor electrodynamics, fact that can be understood thinking to the *Poynting* energy vectors. Each of the *Aries* magnetic dipoles etched into *Zeus* rotor, having an own auxiliary, are reoriented under the greater outer **Hera stator permanent magnetic field** (B), as many sperms of the God flowing into the only empty ovule of the Goddess, depending on their vacuum magnetic

$$\text{/ permeability } (\mu_0) \text{ as following:} \sum_{i=1}^{n} \mu_0 \quad H \overline{\quad\quad} \sum$$

Where:

$4\pi \cdot 10^{-7} \, H \, m \approx 12.57 \cdot 10^{7} H \, m$
H = *Zeus* rotor induced magnetic fields [T]
n = 7 or 12

According the **Lorentz law**, the right *electron force* is done by the cross product of the azimuthal/ velocity and the axial magnetic flux: $F = v.B = v.0.04T = 0.3N = v = 0.3N.0.04T = 0.75 \, m \, s$ that let them helically roto-traslating in a 3D Hilbert space. The electrons accomplish a conical spiral rotating from the border of the real disk to the imaginary hyperbolic Dirac cone, having main axis the *Zeus* shaft of the generator, in a complex counter-clockwise (-) pushing or clockwise (+) descending pulling centered in the singularity of the *Anahatha chakra*, the human heart. *Iridium* (Ir) is the element with the highest oxidation state (+ 7), that means that when linked to other elements like *silicon* (Si) into multilayer film deposited *iridium silicon* (Ir-Si), it can release the highest electron content almost at speed of light, hence, potentially generating the highest electricity values by Albert Einstein photo-electric effect. *Mercury* coils in Ir-Si enveloping tubes would solidify when the temperature would reach (T_{Hg} = - 38 ° C), copper (Cu) seven conductive *Zeus* wives, Poseidon golden (Au) or *Yttrim Bariun Copper Oxide* (YBaCuO$_2$) super-conductive coils would act similarly *Zeus singularity* would be characterized by the inverse reduction potential of the Helmholtz cavity resonator, but with a similar polygonal arrangement. *ElectroMagnetic* (EM) radiation transmitted through the *Zeus* singularity would induce the dipole rotation exothermically in the cavity, smaller compared with wavelength (RPM < λ e.g. infrasound 240 min^{-1} < 0.23 nm), that vary from maximum to minimum amplitude and *vice versa* in 21/billion times/s heating it up, raising the enthalpy, but ⌀sealing it in the shielding electromagnetic shell being characterized by the diameter **pores** insulator = 3.8 nm ≤ 2 Max **amplitude** 2A$_{max}$.

Roto-levitating spherical Planck radiant *Zeus singularity*, done of an inner of reflective superconductive perfect diamagnetic *Hermes* mercury (Hg), characterized by relative permeability lesser than one (μ_{Hg} < 1), immersed in UHV would be trafficked by the magnetic field lines, but enveloped by them would be constrained in the centre, hence, levitating.

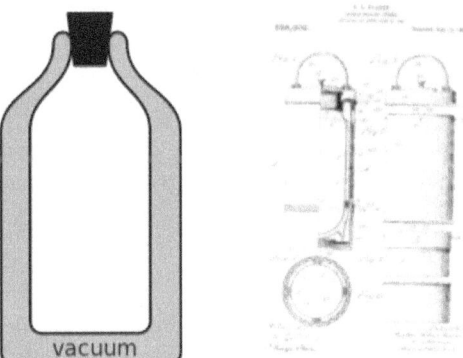

Figure 1 Gustav Robert Paalen, double walled vessel Patent June 27 1908, published July 13th 1909 [contemporary model (left), original sketch (right)]

The outer thermal vacuum insulation (*'Thermos'* invented by James Dewar in 1892), could be done of brass alloy (Cu + Zn), aluminum (Al), borosilicate glass (B_2O_3), bio-plastic, (*'Orion synthetic meat culture'* 1.40 € https://www.amazon.com/dp/B08DDD3Q58) or Ir-Si-aerographene ($SPL_{graphene}$ = 53 dB), *Klaus Halbach array* assembled with typical half-equator strong outside/weak inside *Ares* permanent magnets and the other half, *vice versa*, compensating each other, for focusing particle accelerator iso-beams that, otherwise, would move omni-directionally, or delimited by the walls of the imaginary Bi-Dirac cone in a complex motion.

Material	RSI [m · K/W]
Hermes vacuum insulated	7
Silica aerogel	1.76

The volume of *Hera* in the core of the *Zeus* singularity ($V_{singularity}$ = 70 mL $\vec{r}_{singularity}$ = $\sqrt[3]{\frac{V_{singularity}}{\frac{4}{3}\pi}} = \sqrt[3]{\frac{70mL}{4}}$ = 2.6 cm) is equal to the volume of the other wives solenoids. Being the thickness of the monolayer of graphene w = 335 pm, the radius of the UHV would be equal to: $r_{singularity} - r_{graphene}$ = 2.6 cm − 335 pm ≈ 2.6 cm. That's for which telepathic communication can occur even if the distance is foldable by the Bubble Universe, and the lovers are physically one into another but far field. Nevertheless, the volume of *Hera* globular helical core inflated into the enveloping insulation of the husband, would be equal to the volume of the other tubular solenoidal wives. Finally, the UHV filled by the previous generation of *Zeus*, the baby *Hermes* rototraslating when hot enough in the womb of the mother *Maia* would be done by the difference between the Ir-Si aerographene and *Hera*. The six wives are placed in a configuration all along the stator as in a double Y transformer transferring the potential among them when induce by the permanent magnet etched onto the Zeus singularity.

Many faces would have the father and all of them would attract, but due to the quickly spin they would never be caught, so the JJ locking between *wives* superconductive copper (Cu) coils enveloping the *Zeus* neodymium magnets would inevitably anchor temporarily to all of the wife *Demeter, Eurynome, Metis, Leto, Themis, Mnemosyne, Hera,* choosing the last as reference for the offspring e-devices. *Zeus* rotor surrounded of seven tremolo stators' coils, framed in a circle or in a polygon like a hexagon inscribed in itself, would be the geometrical configuration on which the physics of motors lays for which their variability would depend on the shapes, sizes and material used.

The **six Zeus wives' coils** would be characterized by the following inner to outer radiuses:
1.   Inner neodymium core $l_{NdFe14B} = 1.25$ cm
2.   Conductive copper (^{63}Cu) $r_{Cu} = 1.35$ cm

Hence, by an overall solenoidal cylinder volume $V_{wife} = 7$ cm^3, and its electronics physical quantities are: $N = 70$ turns, $l = 1.25$ cm, $r = 1.35$ cm, $I = 1.8 \cdot 10^{-9}$ A, $\mu = 132 \cdot 10^{-6}$ H/m, $B = 0.0013$ T, $L = 0.000096$ H.

As the heart do not listen sound $[20 < v_{AF} < 20 \cdot 10_3$ Hz (17 m $< \lambda < 17$ mm)], but feel in the senses' control (Sanskrit: प्रत्याहार *Pratyāhāra*) so the *Zeus* singularity would let flow through it just the whispering infrasound of *Hera*. Ultrasounds of the angry *Juno* when discovering *Jupiter* thinking someone's would let emit light from the collapsed bubble singularity of the 'I' of the husband, the liquid mercury, according the *sonoluminescence* phenomenon.

Absorbing material: $\alpha = 1^{-|(\frac{p1}{p2})^{--}e} \ jkx \ ^{(ejkx} + h^{)|} 2\frac{I_R}{h} \propto \alpha = 1^{-} I_R / h \ 0 < \alpha < 1 ^{]}$

  Where:
j = imaginary unit
p = pressure [atm]
1, 2 = sound wave emitters
R = reflected [W/m^2]
I = incident [W/m^2]
k = wave number [m^{-1}]
x = distance between emitters [m]
Exempli gratia, coefficient of absorption of w = 5 cm slag wool or glass silk $\alpha = 0.85$. $\alpha_{Graphene} = 2.3$ %.

Air permeability along the walls vary depending on the following formula:

$$Porosity = \varphi = \frac{V_o}{V_a} = \frac{V_{UHV}}{V_{Hera}} = \frac{\frac{4}{3}\pi r^3_{UHV}}{\frac{4}{3}\pi r^3_{Hera}} = \frac{wV'}{A\Delta p} \approx 1$$

Where:
V = volume [m^3]
o = void
a = absorber
A = area [m^2]
w = thickness [m]
V = volume [m^3/h]
Δp = pressure [Pa]

2 Max **amplitude** $2A_{max} \geq$ ⌀ diameter **pores** insulator = 3.8 $\int$nm. Tortuosity (q = Φ R$_s$/R$_f$), flow

resistivity ($\sigma = \Delta P/Ud$), viscous characteristic length ($\Lambda = 2\frac{\int v^2_{fluid} dV}{\int Sv^2_{fluid} dS}$) and thermal characteristic

length ($\Lambda' = 2\frac{\int dV}{\int dS}$).

The **outward radial pressure** of the *Zeus* rotor singularity magnetic field due to the neodymium magnets ($Nd_2Fe_{14}B$) according the *Meissner effect*, of the Ir-Si aero-graphene complex solenoidal (real empty tube with imaginary null loop distance) super-conductive copper metameres of the torus *Hera* stator would be null when two opposite inductor *wives* coils are active:

$$p_{mag} = \frac{B^2}{2\mu_0} = \frac{16 \cdot 10^{-3}}{25 \cdot 10^7 \frac{H}{m}} = 0 atm$$

The strongest *Zeus* permanent **magnets** in the world, placed according the *K. Halbach* array, but along a circumpherence is composed of an alloy of neodymium, iron and boron ($Nd_2Fe_{14}B$) structure usable for *Hera* stator and *Zeus* rotor's coils magnets of and three *Hera* voice-coils. The rotating pattern of *Aries* permanent magnets (on the front face, left, up, right, down) can be continued indefinitely and have the same *magnetic field augmentation* effect discovered by John C. Mallinson in 1973.

A magnet can lift the singularity of *mass* (m_s) according the following derived *Maxwell equation*:

$$m_s = \frac{B^2 A}{2\mu_0} = \frac{16 \cdot 10}{25 \cdot 10^7 \cdot 1 \cdot 9 \frac{H}{m}} \cdot n 71^9 mm/s^2 = 1.2 kg$$

The 2D *Aries magnetic attractive or repulsive ideal field force* capable to suspend a magnet (maglev) on the active side of the K. Halbach half equator as planet in the free-space would be equal to:

$$F_{(xy)} = Fe^{ikx}e^{-ky} = \mu q^+ \frac{q^-}{\pi r} s = 0.3N$$

Where:
x, y = coordinate [m]
k = wavenumber or repetency [m^{-1}]
μ = magnetic permeability = 1 T ·
m/A q = charge [A/m]
m = mass [kg]

The *Zeus* singularity in the centre would describe 7 loops moving from the centre O to 1, than thanks the centrifugal attractive force and the activated opposite wives coils, it would spin and move to 2 place at $\alpha = 180°$, than $\beta = 90°$ to 3, $\gamma = 180°$ to 4 and so on till 7 making an imaginary sacral *Flower of life* with its harmonic motion centered in O, following the pendulum motion laws. The attraction from a fixed strength *Aries* magnet decreases with increased distance, and increases at closer distance according the *Lenz's law*.

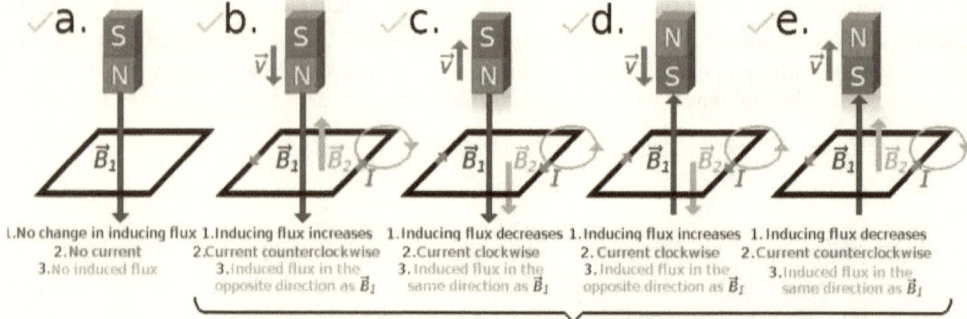

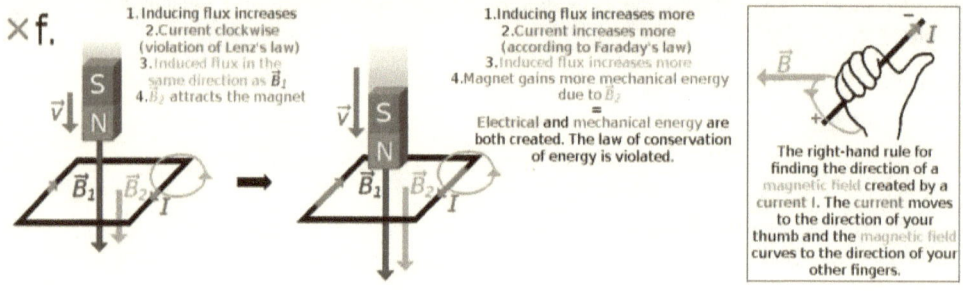

Figure 2 Lenz's law diagrams

The signal processing *Hilbert transform* a function u(t) into another H(u)(t) shaping itself is done by

$$H(u) = \frac{1}{\pi} \int_{-\infty}^{+\infty} \frac{u}{t-\tau}\, \tau\, d\tau$$

Generator mode

Just imagine your heart magnetic field, where *Zeus* and *Hera* divine royal are in love, and the opposite direction charged vortexes (q^+ and q^- - e.g. electron and protons) Dirac bi-cones rotation in, one real the other imaginary in a complex mechanical motion surrounded by the *Manas Sharir* firmly shielding interferences in the **AC alternators** leaving the magnetic field stationary within the conductive propeller, otherwise, spinning coherently in a roto-translation that would change together with a variable degree of synchronism with the cones in which the *Zeus* shafts anchored onto the generative singularity in the **AC/DC dynamos**, in both cases converting the mechanical energy from the renewable fluids isolated from natural environments like daily *Artemins* tides and seasonal rain and *Aeolus* winds, or the helically flowing onto the coppery *Zeus* wives spirals for triggering the impeller(s) turbine rotation cooling **MagnetoHydroDynamic** (MHD), initiated by the Nobel Prize in Physics in 1970 Hannes Alfvén, effect called *self-focusing*, a magnetic property of an electrically conducting fluid even the hyperspace of the *Solar System* (SS) in which it would induce magnetic field that in turn polarizes the medium and reciprocally changes the magnetic field itself generation electricity (I) in external circuits according the Faraday law of induction. Otherwise, the atmosphere between *Hera* stator and *Zeus* rotor could be maintained under isolated *Ultra High Vacuum* (UHV) condition were the molecular content would be regulated by innovative,

'Zeus Genset' (generator and engine) pag. 6

but ancestral **Solar** a mixture of *Helium* (He) and *Hydrogen* (H) favouring the up thrust of the central *Hera* stator, increasing the electricity generated, but reducing its the durability. The gaseous bubbles under UHV, immerse in non-corrosive liquid medium (e.g. peracid, alkyl carboxylic acids and peroxide) where looping acoustic standing waves flow would emit a flashing pulse of light with each compression on the rhythm of the heartbeat, the *Hera* globular coil should let turn the wave within itself emitting electromagnetic wave ($P \approx 5$ mW for $t = 50$ ps) with a more stable period and position than the originary oscillatory heart-like sound wave. The *Zeus* rotor of the genset would levitate beigns lighter than the surrounding noble gaseous inflated into the UHV medium, counteracting gravity providing a stronger and opposite ElectroMagnetic Lorentz force, in a stable position in the centre of the *Hera* stator done of a segmenting toroidal inductor the outer electrical *Poseidon* conductive shell for leaving the *Ares* magnet metamerically in contact with the air gap drawing two side circumpherences for each *Hera* stator coil wished, precisely, seven *Zeus* wives. Fluid with zero viscosity and null entropy, flowing without friction, is in a state of **super-fluidity** discovered by Pyotr Kapitsa, John F. Allen and Don Misener in 1937. He-3 than He-4 is easier to maintain itself in a superfluid state increasing its temperature, but both reached this state of mater can be routed by Nirvanic Johnson supercurrent and quantized vortex lines and rings done of the sub-atomic particle's orbital deflections. ^{4}He is solid just above pressure $p = 25$ atm, fluid ^{4}He-I and superfluid ^{4}He-II are separated by the λ-line in the point at temperature $T = 2.172$ K, pragmatically separable inserting a floating smaller Becher in the bigger containing them together, observing the faster superficial superfluid He-II driven by surface tension and for capillarity raising the walls of the smaller Becher filling it. The *instantaneous force* that drives the motion of the superfluid ^{4}He given by Isaac Newton:

$$F_{He} = m \, dv/dt = - \nabla (\mu + m \, g \, z)$$

Where:
m = mass [kg]
v = velocity [m/s]
t = time [s]
g = gravitational acceleration [m/s^2]
μ = molar chemical potential
z = vertical coordinate

^{4}He can be seen in term of *Bose-Einstein Condensate* (BEC) acting as photon bosons, while, *Bardeen-Coopers-Schrieffer* (BCS) super-conductors as describe their smaller isotopes ^{3}He fermions interacting via spin pairing. BEC and BCS extreme states can cross themselves in some ultra-cold gaseous clouds as the ones characterizing the hyperspace during the Ice Ages on Earth. Nevertheless, the superfluid can be applied as quantum solvents, gyroscopes fillers predicting gravity responses variable, for example, with altitude, or as cryocoolers. Four distinctive Multi-Universe heterogeneous globules make the original spheroids with a liquid mercury surface projecting in its Solar (H + He) noble gaseous core four coordinative images, each one completely reproducing the source shapes of the massive particle impressed, instantaneous reflected by the diffuse light interfering with the bubble dynamic rotation along the time line. Superfluid helium drops He$_{(l)}$ at temperature $T = 0$ K resolving the hydrogens H$_{(g)}$ acting as quantum entanglement *'Einsteinian spooky particles at distance'* manifesting their instantaneous localizations, not in two opposite far-off, but in four quantum states coherent with the Cartesian Coordinate System in the Multi-Universe domain.

POLAR P 67104 2T P = 800 W
(https://www.google.com/shopping/product/13093787224963482952?lsf=seller:101019159,store:1
4016738610532907887,s:h&prds=oid:6478442582080073106&q=genset+buy&hl=en&ei=aYv5Xp
mHCMStrgTF3oyIBg 90 €).

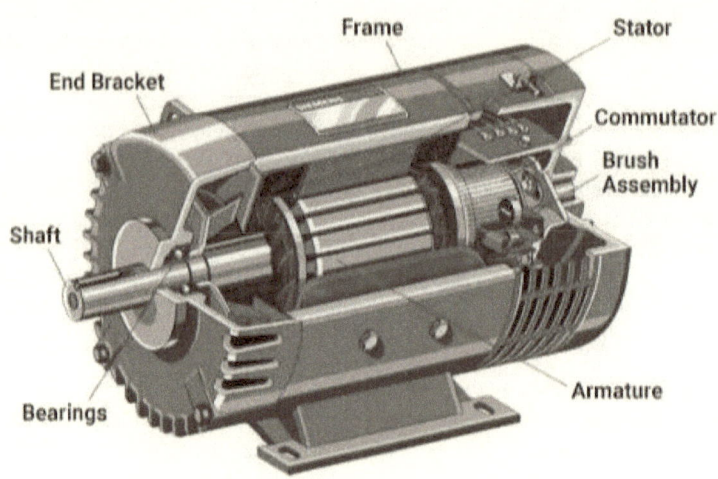

Figure 3 Generator - Image Source: https://www.electricgeneratorsdirect.com/stories/1485-How-
Generators-Work.html

Fixed the bearing relationship each other, the coupling between the Gods would let reach unsustainable temperature when *Zeus* rotor oscillates within *Hera* stator, in a noise condition that just the smoothing elevating medium would scatter, whenever the armature would not condensate enough the heat into it. The load on the motor is frictional rather than inertial as the friction reduces any unwanted oscillations. The stability of the magnetic dipole *Zeus* rotor wouldn't be accomplished when the motor is switched off according the *Earnshaw's theorem*, but, certainly, wouldn't be a problem if minimum 2 opposite *wives* coils would be maintained in a background levitating mode, that guarantee the *Zeus* singularity rotor to do not shift too quickly onto *Hera* stator breaking it when oscillating within it. Hence, a trade-off between the emptiness of the gap and the RPM, the toughness, the friction according the Newtonian motion's laws would variably determine the resilience and performance of the device. The time would be determined by the empty space. <
< How long we can stay here alone? *Hera* would have been thought that time when *Chronus* were not looking for *Zeus* for regulating the mortals on Earth >>. The deception of *Zeus* would have been accomplished not in the sky where he was working throwing lightening scaring *Europa* and other plenty of other mortals, but in the ocean, where the electrolytic water would clean, calm, an heal the gonad pregnant of *Aries*. A thin (w ≈ 2.5 nm) relatively cheap aluminic oxide (AlO_2 x nH_2O) anodizing passivate layer (I_{DC}/A ≈ 150 A/m², ΔV ≈ 150 V, P = 22.5 kW) dissolving nitric acid
(HNO_3) in a m_{Na3PO4} = 60 g : 45 L H_2O [$\rho_{\frac{HNO_3}{NaPO_4}}$ = 1.5 g L   /] for t = 12 h, supplying (DC 12 or AC
230 V), placed for preserving the thrustful of the shaft of *Zeus,* and reducing to the minimum the power losses, proportioning the in and out intensity with each cycle.

In sensu lato she generates and he acts as motor, but when they pair were in one only genset arranged logarithmically continuous vertically for mechanical transduction or pulsing horizontally for generating electricity and varying the relative velocity between *Zeus* rotor and *Hera* stator (v = 4
'Zeus Genset' (generator and engine) pag. 8

ZZ   Thrust bearing spinning velocity of the *Zeus*(rotor increases) proportionally with the amount of electrical current flowing through the generator
Ɒ Ꝓ Ʋ ɑ 7.Hz, SPL = 120 dB, RPM = 240 min , P = 1.5 W, ΔV = 3.6 V, R = 10 Ω, I = 0.36 A). 608

Figure 4 *'Jupiter and Juno on Mount Ida'* by James Barry.

Perhaps, the deity had endless love and with uncountable watts, through the waves compared to the mortals that shaking via foot pumping, hand cranking, or whatever other muscle engagement a steady power P = 75 W during t = 8 h, while, a 'first class athlete' can produce approximately P = 300 W for the same period. At the end of which an undetermined period of rest and recovery will be required, leaving the normal human becomes exhausted within just 10 min.

Homopolar DC Faraday generator, inspired by astrophysical objects like *Uranus* aurorae, black holes coalescence and binary stars, are done of rotating conductive flywheel perpendicularly to a core shaft enveloped by static magnetic field with a different potentials between them, filled into inert noble medium, into which the *Zeus* plasma sparks diffuse, could generate magnetic field up to B = 0.8 mT orbiting on a common barycentre. The invoker of the *Caelus* would evoke potential over their *Brain Blood Barrier* (BBB) enlightening though the pineal gland the telepathic communication in the *Brahma Sharir* between two far field *Homo sapiens* and the indie *Homo atm* capable to control telepathy in a simple way, when thoughts become visible when are not yours anymore, and your brain works as a unipolar generator for managing the more susceptible to the relatives mind ascribable to a MHD generator.

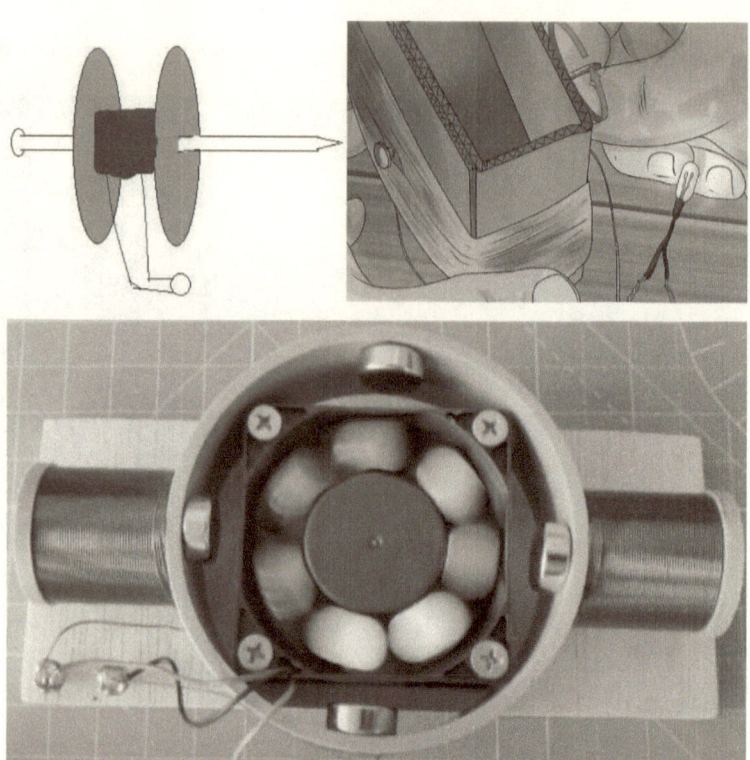

Figure 5 *Free energy generator* enlightening a *Light Emitting Diode* [(LED) P = 5 W, ΔV = 220 V, I = P/ΔV = 2.3 mA, T = 6500 K] done of a nail (l = 15 cm), two cardboards, insulator and copper wire (N_{Cu} = 1500, w = 0.287 mm) (top) or a *fluorescent bulb* (P = 60 W, ΔV = 220 V, I = P/ΔV = 272 mA, T = 6500 K) (bottom) : wood base, E14 bulb holder, two parallel screws with eight bolts, tapeø (w_{Cu} ≈ 1 mm), two insulated copper wire (l_{Cu} ≈ 7 cm each), Neodymium ($^{60}NdF_{14}B$) *Zeus* ring magnet (5 < < 25 cm, 12 · 4 mm N35 2 €) cyano-

acrylate glued on a fun, pliers, abrasive paper, wax, aluminium contacts soldered. P.S. Electrolytic light free energy bulb can be made soldering Zn and Cu electro-strings onto it and submerge into a NaCl solution – Image source : https://www.instructables.com/id/how-to-make-the-simplest-electric-generator/

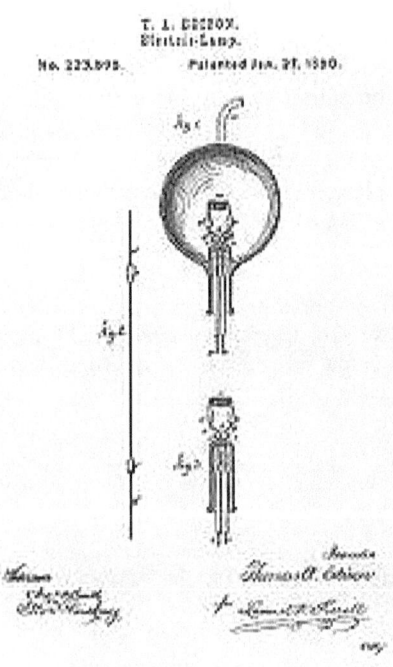

Figure 6 Electric-Lamp (U.S. Patent#223898 - 27th January of 1880) Long lasting *incandescence bulb* (Edison standard Screw 5 < ES < 40 mm), invented by Thomas Alva **Edison**. Among the other creations of this advantageous American inventor there are: the quadruplex telegraph (U.S. Patent 90.646, granted on 1st June of 1869), the phonograph, the motion picture camera, the electric vote recorder, the calcium tungstate fluoroscope, a machine that uses X-rays to take radiographs, brighter than the *PlatinoCyaNide* (PtCN) previous made by the German *Wilhelm* Conrad *Röntgen*, carbon condenser microphone (two metal plates stabilizing the carbon corpuscles in between that increasing the amplitude of the acoustic wave flowing within it raise the voice pick), the *tasimeter* InfraRed (IR) measurement tool, the stock thicker, the first electricity-based broadcast system – Kinetoscope, and phenol (aromatic ring + OH) production, perhaps, for resonance purposes.

Inspired by the white hole side of the wormhole, the following bulb would emit energy-matter when *Uranus* laser light pulsing beam polarized onto the *'Zeus singularity'* bubble ergodically describing refraction pathways against the Ir-Si-aerographene porous enveloping spherical walls, following the *Snell's law* with *Refractive Index* (RI) of the noble helium hexagonal atoms ($n^2_4He = c/v = 1.000034912$, where c = speed of light in the vacuum = $3 \cdot 10^8$ m/s and phase velocity in the full medium $v = \lambda/T$ = wavelenght on time period https://refractiveindex.info/), hence, sealing into the *Solar* cryocooling clouds cyclotrons exited to a higher quantum state (LUMO -> HOMO) stripping-out the orbitals and leaping, accelerated and collision in chains reactions (electron avalanches) generating plasma current (LIPC) inside the bubble, finally, glowing the **fluorescent bulb** of sequential seven main distinctive mode colours according the visible light spectra (PAR) photoelectrically due to related to the *Zues* wives coils of distinct nuances depending of the localization along the *Seed of life* (seven wives Zeus deception series: 1. Demeter, 2. Mnemosyne, 3. Eurynome, 4. Themis, 5. Metis, 6. Leto, 7. Hera), a rainbow of colours sanctioning the birth of *Ares* after the thunderstorm due to the resonating coupling between *Zeus* rotor and *Hera* hexagonal stator. Is to note that the design of the stator has to be done according the crystal structure of the medium containing it as quantum mechanics which being the observer within the medium observed, the electricity will flow into the coils etched in the hexa-ring in a continuous flow, despite the attraction-repulsion between *Zeus* and the wives, and their wish to proliferate is discontinuous, ending only with the marriage of the levitation with the globular coil into it, in other words, when the electrical current would displace from anode (+) to cathode (-) of *Hera* coil, slightly depressed, scattering the electrons into the *Solar* medium into sparking lightenings properly insulated in the bed of the divine deception.

The coupling factor and the delay should have been considered for evaluating the resonant natural frequency of the superconductive self-generated magnetic field Singularity quantum levitating and locked would store great amount of power as electric current (Boaz Almog, TED talk https://www.ted.com/talks/boaz_almog_levitates_a_superconductor?language=it).

The *Zeus Singularity*, could become biotechnological, adding continuous glowing *luciferins*, isolated from the aqueous cytosol lantern cells of the **bioluminescent** organism's firefly *Photinus pyralis*, reacting with catalysing enzymes *luciferases* or recombinant *Escherichia coli* plasmid bringing their *CoDing Sequences* (CDS), having as substrates Oxygen (O_2) and *Adenosine TriPhosphate* (ATP), or the photo-proteins Ca^{2+}-requiring *aequorin*. *Luciferins* in seven syringes on the hexagonal ring, while, *luciferase* into a central globular silicon bioreactor substituting theⱺ superconductive coils with the enzymes, let pierce and fill the Singularity with a capillary needle (
= 38 nm) experiencing *Zeus* after each infertile mating with sinusoidal photo-electric waves.

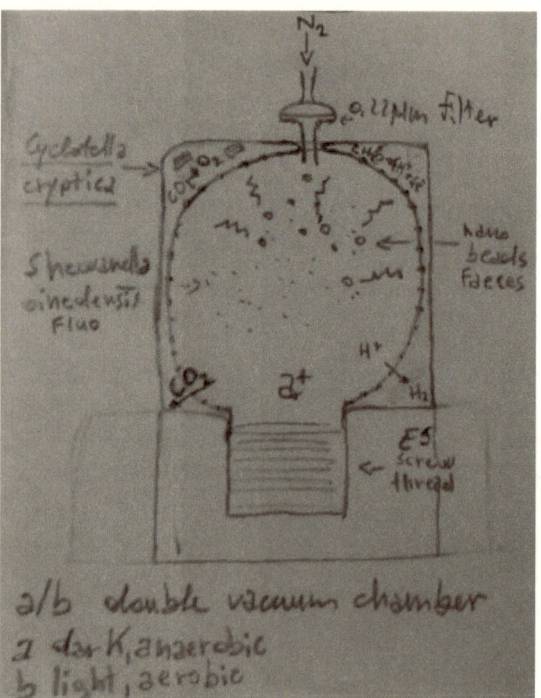

Figure 7 Deductive sketch of globular bulb *Microbial Teleportation Cell* (MTC) chamber for in series gravimetric bioreactor (for more info about *'Illusion and reality about teleportation'* © Copyright Antonio Silvestro, 202x).

Piezo-floor generator

Bottom and upper plate, four springs, one bigger central or four smaller side piezos, set of gears fitted on rack, generator, battery (ΔV = 9 V or LiPo) black (-) / red (+) alligators and leads, diodes, MOSFET, capacitor, Arduino UNO Wi-Fi ESP 8266 (10 < P_{step} < 20 W) https://www.energy-floor.com, https://www.pavegen.com.

Bike dynamo generator

A bicycle generator is done adding a back wheel without black plastic tire and an elevating support, DC motor (RPM = 2900 min^{-1}, P_{out} = 350 W, 12 < ΔV < 24 V, I = 19 A, 50 €), motor and battery leads (4 €/each), battery charger (P = 300 W, I = 20 A), inverter DC/AC (P = 400 W, 25 €), LCD, diodes (1 €/each), pulley (ø = 6 cm), transmission chain-belt (5 €), charge controller (I = 30 A), lead (Pb)-acid battery (12 < ΔV < 48 V, I = 18 A, 35 €), (ΔV =2.5 V), switch and LED => P $_{out}$ ≈ 1.5 W/cycling 10 min/day suitable for electro-domestic such as light bulbs, grinders, fans and phone battery (C = 1 Ah, V = 12 V, I = 2 A, P = 24 W, $t_{cycling}$ = 7 h).

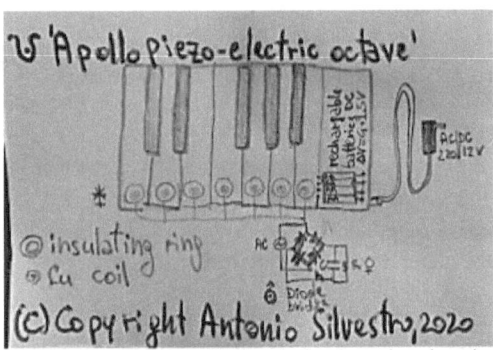

Figure 8 'Apollo piezo-electric generator octave' sketch, keyboard recharging in a feedback loop while playing music. Each of Jupiter piezoelectric crystal (V = 1 mL, F = 1 N), of the electric piano keys (or also pads of an e-drum) would generate Alternated Current (AC) that would flow into the gate of the Mars transistors collected into the Uranus TRIAC or diode bridge rectifying it into Direct Current (I_{DC} = 12 V) passing through Venus resistor, Saturn capacitor filtering the noise out of the circuit enlightening a light bulb. The AC/DC transformer so built would make the Apollo instrument able to bright the chamber in which the invoker musician is playing not just with sound, but also with light in a sonoluminescence effect phenomenon - Image source: © Copyright Antonio Silvestro, 2020.

Motor mode

The **engine** (Latin: *ingenium* 'ingenious') is a machine aimed to convert different kind of energy like heat, electricity, pressure, elastic, chemical into mechanical energy. Engines have ranged from 1 < cylinder < 2^4 = 16 designs with corresponding differences in overall size, weight, engine displacement, and cylinder bores. 4 cylinders and power ratings from 19 < P < 120 Hp : 14 < P < 90 kW. 1 cylinder is 3000-folding the power of the human heart, so being the natural highest velocity about v = 10 m/s, the mechanical device would run at v = 3 km/s if the proportionality wouldn't be exponential, but linear, ideally being the strength the same), but the average mass of this last m = 240 kg, three-fold the once of the average adult human, its acceleration should be thrice for equivalating the motor according the following equation:

$$3a_{human} = a_{motor}$$

Air-breathing motor are mainly classified into steam, pistons reciprocating, ram, centrifugal or axial compressed turbine jetting propelled and detonating wave pulsed. The *chemical composition* of the exhaust gas produced by the sparking ignition is the following: N 70 %, H O 10 %, CO 10 %, CO 5 %, H_2 3 %, O_2 2 %. *Hephaestus* spark plug triggering the burning of compressed fuel has a metal shell ceramic porcelain insulated ($10 < < 20$ mm, $0.095 < 1 < 2.649$ cm), commonly heat-pressed sintered alumina (Al_2O_3) obtainable in a kiln at $T \approx 1300$ °C, white glazed on the top while not light reflecting at bottom, withstanding $500° < T < 800°$ C and $\Delta V = 60$ kV, where is screwed around a long-lasting sharp (e.g. iridium ^{77}Ir or platinum ^{78}Pt), theoretically pointed (or pragmatically for disposable plugs), central superconductive *Mercury* cathode (-), to which may be applied a heated resistor for increasing the thermionic electron emission, joined with a high tension ($\Delta V = 45$ kV) pulsing ignition magneto, transformer, LC circuit and/or other electric current generators. Certainly, the thermo-electrical insulation can be made of other materials resisting the same high temperatures and ElectroMagnetic interferences like the phyllo-silicates (mica) that eventually can be passivated for reducing oxidation.

On the contact imaginary segment spark and surrounding gas injected, depend the ionization happening, *exempli gratia*, air, having a dielectric strength of 3 MV/m, would be involved in the molecular dissociation by the plug just if at distances $1\ 10^{-4}$ m at maximum voltage mentioned. The electric current flowing through the ignition *wives* coil into the plug channel would reach the high temperature of $T = 6 \cdot 10^4$ K and appear similarly to *Zeus* lightening.

In *External Combustion* (EC) engines, a capacitative heated exchanger ignites flowing fluid, while, in *Internal Combustion* (IC) the expansion of the high temperature and pressure gases combusted applies force to the engine's components such as pistons, turbine blades or nozzle, moving the motor over a distance via the generation of mechanical work.

Four-stroke combustion engine cycle in the *Hephaestus* chamber:
1) Induction (fuel enters)
2) Compression
3) Ignition by burning the fuel with a spark - complete combustion: $CxHy(l) + zO_{2(g)} \rightarrow CO_{2(g)} + zH_2O(l)$
4) Emission (exhaust out)

The smallest *biological motor* is the *myosin actin-binding that via ATP hydrolysis*, in the sarcomeres of the *'Orion synthetic meat'* make them in contraction-relaxation duality that force the motion. The vertical motion of a piston within cylinder could be governed by a mechanism in a superior scale that mimic the muscle contraction being the walls of the cylinder done of actin and the core of simplified myosin piston. This last done of two elastic entwining helixes that bonding with their head hooks successive teeth in the internal actin, tropomyosin, troponin walls, when the ATP_{ase} cleave its substrate (ATP = 1€/g) releasing energy for the myosin-actin cross bridge binding cycle, hence, the piston sliding. As coordinated rowing, the myosin -mers paddle through the waves of the chemical fuel ATP hydro-solution.

Myosin steps:

1) Rigor conformation: myosin and actin bond

2) ATP bind myosin
3) ATP hydrolysis into ADP and Pi
4) Power stroke: Pi releasing

Hybrid myosin *Internal Combustion* (IC) motor:
1) Energetic powder fuel sprayed in the actinous cylinder
2) Fuel bind myosin inducing the compression of the chamber with the filament's contraction induced
3) Chemical reaction that break down the reactive fuel in two products
4) Releasing of the by-product

Biosynthetic motors have been created with *Zeus* rotors like the triptycene turning within the helicene thanks substitutive reactions involving functional groups changing onto aromatic rings of three $\alpha = 120°$ angulated propellers, from amine (-NH$_2$), through isocyanate (R-N=C=O), to carbamate (-NH$_2$COOH) (T. R. Kelly et al., 1999). Pyridine-based catalyst promotes carbonate-forming irreversible reactions that ratchet the displacement of the macrocycle chemical fuelled by 9-FluorEnylMethOxyCarbonylChloride (David A. Leigh et al., 2016). Aryl: Naphthalene (Feringa et al., 2005) and Aryl-Biphenyl lactone (Branchaud et at., 2005) *Zeus* rotor : *Hera* stator would be other torques of powerful chemical motors of synthesis. Thermal helical inversion (t = 0.01 s) of methyl groups involved in isomerization trans-cis of (P, P) light-driven characterize the 9H-Fluorene obtainable via 'Barton-Kellogg reaction'. In parallel LC resonating nano-biosynthetic engines could bring to macroscale extremely power also wirelessly.

Electric motors, firstly invented by Moritz Jacobi in 1834, convert electric current into motive mechanical, powering energy both by switching commutator *Direct Current* (DC), such as from batteries, motor vehicles or diode, transistors rectifiers, or by (super)-conductor, brushed or brushless, asynchronous or synchronous, fluid cooled *Alternating Current* (AC) sources, created by Walter Baily in 1879, such as a power grid, inverters or electrical inductive generators, and asynchronous motors. The main components common to these electric devices are the moving *Zeus* rotor with conductive copper (Cu) or aluminium (Al) winding wraps around a *Aries* ferromagnetic (Fe) core forming poles, the firm magnetic Neodymium-Iron-Boron alloy (^{60}Nd, ^{26}Fe, ^{5}B) *Hera* stator slot, the supportive bearing for holding in place the *Zeus* shaft, the minimal air-vacuum gap possible for avoiding friction. The originary shaded-pole motor was like a Faraday transformer done of *Aries* magnetic ring and inductive winding *wives* coil with an accessory *Zeus* shaft supported into it perpendicularly rotating, as magnetic and electricity field. *BrushLessDirect Current* (BLDC) can reach P = 100 kW are applied from CD/DVD to aircrafts. Universal AC/DC are lightweight and relatively fast speed v = $3 \cdot 10^4$ RPM are used instead for portable working and electro-domestics and tools. Precise and smooth differential rotation of stepper motor make them stopping in instants and are usually utilized for digitally and optical devices. Performance of the asynchronous *Alternating Current* (AC) induction motors increasing the cross-sectional conductive area, overall aerodynamics, reducing the noisy vibration of the supports and the friction related with silencing mufflers.

Thrust force happen between *Zeus* rotor and *Hera* stator, while, **torque** is the energy generated by forcing a lever of length from to a shaft (reducing) (as rotation) catch able via the following form $\cdot$ **0.3** $kg\ m\ s^2$

Longer cylinder would be more energetic than shorter using the same force. *Exempli gratia*: windshield wiper motor screwing auger powerful gear GF45 motor, P = 66 W, stall torque τ = 30 m N (v = 0 m/s) and runs at speed v = 35 RPM on the low velocity setting, toothed belts for the X, Y drives and a leadscrew for Z-axis through which poseidonian fluid can flow. Degree of rotation can be assessed by 'Resolver' both analogic and digitals. *Direct Torque Control* (DTC) can indirectly manage the power of *Alternated Current* (AC) engines. When, the corner is right the sin α = sin 90° = 1, so the energy involved is maximum (E_{max}). The engine **power** can be calculating as follow:

$$P = \omega \cdot \tau = 2 \cdot \pi \cdot r = 2 \cdot \pi \cdot r^2 \cdot F = v \cdot F = 35 min^{-1} \cdot 0.3 . \mathrm{V}$$

Where:
ω = angular velocity [rad/s]

Closed-loop *Pulse Width Modulation* (PWM) controller for speeding and torque regulation.

Overall, a motor needs to be simple in built the robust compact size and accurately controlling the high speed, cheap, but reliable with a long lifespan, low maintenance and noise, high in self-starting powerful torque (P = 375 kW) and efficiency, to have electromagnets, precise in turning the *Zeus* shaft.

A **servomotor** is a rotatory or linear actuator, a closed-loop servo-mechanism, which allows for precise controlling of angular or linear position, paired with an encoder for providing velocity and acceleration. The encoder and controller of a servomotor are an additional cost, but they optimise the performance of the overall system compared to the relatively cheaper **stepper motors** (https://www.wish.com/search/stepper%20motor/product/59138fd807797b02eb273868?source=search&position=37&share=web), brushless synchronous AC/DC electric commutator motor that moves the *Zeus* shaft coherently with the digital input pulses in discontinuous square wave discrete steps controlled by the indexer microprocessor, precisely, four per turn (4/turn). In this last motor the central *Aries* magnetic core in the uterus of *Hera* stator, interact with 25 electromagnet teeth on the gear energized by an external microcontroller (25 · 4 = 100 steps per full rotation and each step will be 360/100 = 3.6°). Permanent, hybrid or variable reluctance according the magnets flux continuity between *Zeus* cathode (-) rotor and *Hera* (+) anode stator. The Faraday induction of EM field among *Aries* paramagnet coils induce the rotation of the *Zeus* shaft along the perpendicular axis with the radius of the gear. A part from mechanical vibration, electronic drivers' circuit are the main performance amplifier factor converting the indexer signals into power necessary for energizing the windings of the stepper uranium unipolar 4-phase motor. When controlled by and constant voltage drive (ΔV = constant V), the maximum speed (v_{max}) of a stepper motor is limited by its inductance as the father transformer *Zeus* has seen himself in childhood in its hermetic mono-coil form. To obtain high torque (τ) at high speeds (v) requires a large negative drive voltage [ΔV = V_{Zeus} - V_{Hera} = - - (+) = -] with a low resistance ® and low inductance (L).

Motor resonance formula:

$$v = \frac{100}{2\pi} \sqrt{\frac{2p.M_h}{J_r}}$$

Where:
M_h = energetic torque [N · m]

p = pole pairs

J_r = rotor inertia $[\text{kg} \cdot \text{m}^2]$

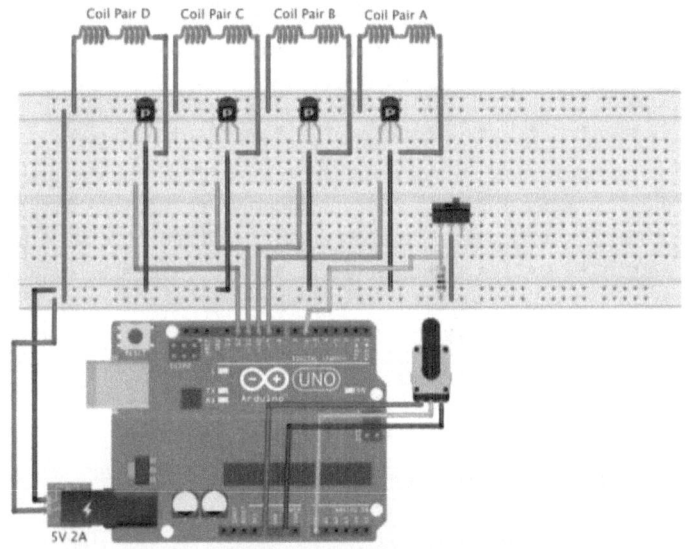

Figure 5 3D printed DC motor: conductive copper wire (l = 1 m) enveloped around each of the *Zeus* permanent Neodymium electromagnet nails (w = 0.3 mm, l = 5 cm 5 % of the coil length) with the drill from tip to head. Inductors so made opposite pared in series (*Demeter-Mnemosyne, Eurynome-Themis, Metis-Leto*) and connected to a breadboard with 4 P-MOSFETs, and to an Arduino Uno. Resistor (R = 1 kΩ) and power executive toggle switch or infinite switch (= simmerstat) usually combined with resistive heating elements. Each harmonic of the radiation emitted produced by the disturbance due to the non-linear loads (e.g. discharge lightening, rectifiers, diodes, transistors, saturated magnetic devices, electrical motors, and the non-ideal transformer), characterized by a diverse frequency can be modulating and differently carried conveying the information. DC power supply (5 < ΔV < 12 V: 1 < I < 3 A) like the USBs. *Ares* and *Zeus*

South poles (S) repel themselves, beigns this last glued on the outer border of *Hera* rotor North (N) facing via a *'Compass Smart'* app. - Image Source: https://www.instructables.com/id/3D-Printed-Stepper-Motor/

Hybrid synchronous stepper motor supplied by 4 separated power sources are combination of *Aries* permanent magnet and variable reluctance, in which the magnets are so many that their surface seems one. Its *Zeus* rotor can be done of 4 successive gears separated by three laminate permanent *Aries* magnet rings of alternate polarity, each offset sprocket counting 25 magnetized cogs; while, it's *Hera* stator of not magnetized internal 42 teeth, 5 for each of the 7 inductors. The alignment step angle of a variable reluctance would be higher α = 40° (wave) or 20° (full) (*half stepping*), compared to the hybrid stepper β°. Three different arch length can be disguise in the attraction-repulsion interaction that conceal 48 cogs *Zeus* rotor and 42 teeth *Hera* stator: aligned, unaligned and half-aligned. Breadboard with two H-bridges done of 4 *Metis* P-MOSFET and 4 N-MOSFET and an Arduino Uno. Wave and full driving both with about 200 steps/cycle, but with one and two coils, respectively, the hybrid half step would be capable of making 400 steps/cycle, 1/16 steps even till 3200 steps/cycle 4988 *MicroStepping IC*, is a kind of advanced H-bridge, that can substitute it in a circuit done of capacitor ©, resistor ®, potentiometer making the motor rotating in a more fluid and less noisy way.

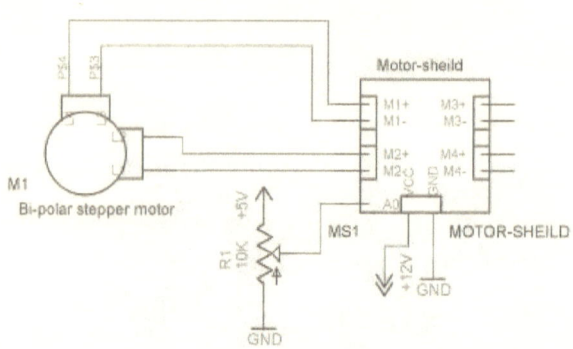

STEPPER MOTOR CONTROLLER FOR SPOOLER

Figure 6 Bi-polar stepper motor spooler schematic – Image Source: Antikvora, 2012

The Albert Einstein theory of **general relativity** suggests that light travels at a constant speed c = 299,792,458 m/s in a vacuum, value coming from the equation of the rest energy: $E_0 = m\ c^2 = p/v\ c^2$ [J], where: m = mass [kg], p = momentum [kg · m/s], v = velocity [m/s] (e.g. ideally a human (*Homo atm*), even an inferior *Homo sapiens* placed in the empty space could reach the speed v_{Human} = 24 · 10^9 m/s). **Perihelion of Mercury** according A. Einstein appear anomalous than the Newtonian ellipse *Seed of Life* like due to the bending of light through gravitational lenses and redshift field, being shifted of a sigma factor:

$$\sigma = \frac{24\pi^3 L^2}{T^2 c^2\ 1-e^2}) \ Rad/revolution]$$

Where:
L = semi-major axis [m]
T = orbital period [s]
c = speed of light = 3 · 10^8 m/s
e = orbital eccentricity

Parallel electrical signal flows via *connexon* hexagonal ion channels of gap junctions done of six connexin proteins that closed would resemble the *Seed of life*, a knot with six loops or the *Zeus* singularity rotor complex path which bisectors correspond to the 7-coloured segment from the centre to the imaginary circumference of the *Hera* stator.

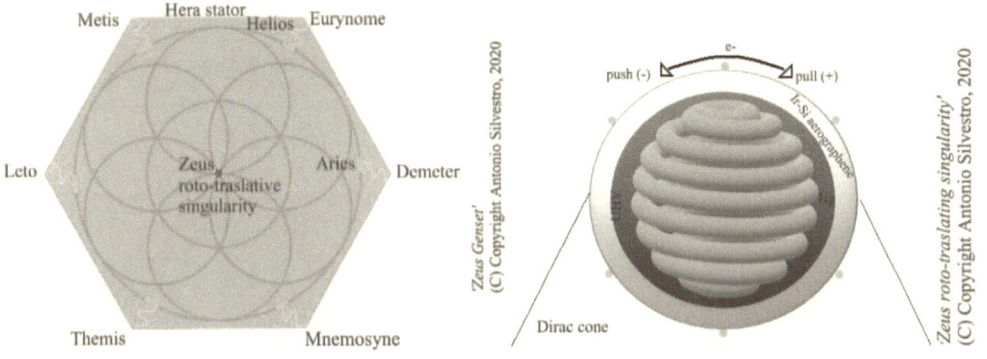

Figure 7 *'Zeus genset'* (3D Builder model One_25mm_hex_3mm_walls.stl) done of a *'roto-traslating levitating singularity'* along the imaginary seed of life counter-clockwise pathway attracted and repelled by the *Zeus* permanent neodymium magnet through the seven wives coils placed in UHV dripped of Solar gaseous solution (H + He) sparking, when, at each passage of the God to the peripheral loved, the medium glows ionize into plasma – Image Source: © Copyright Antonio Silvestro, 2020.

Tachometer

Engine speed is measured by contact or optical laser, *InfraRed* (IR) beam contactless **tachometers** (https://www.instructables.com/id/Measure-RPM-DIY-Portable-Digital-Tachometer/) in *Revolution per Minute* ($1500 < RPM = 60 \cdot 1000$ / (millis() - time) $\cdot REV/2 < 20000$ min^{-1}, v = 360°/min) done of sensor (two resistors $R_{1,2} = 270$, $33 \cdot 10^3$ Ω, 10^3 sensitivity regulator potentiometer, sanded and flattened IR photodiode and IR LED enveloped together in black paper strip and soldered (1. shorter – lead PD with longer + LED), status PAR LCD (16 x 2 cm) with $R_3 = 270$ Ω soldered, triple ribbon cable soldered to the remain leads (2. + PD, 3. LED), perfboard and headers (GND, VVC, signal), 74HC595 8-bit shift register interface using just 3 data pins instead of 6 in which LED backlight is controlled by a N-MOSFET connect to the QF pin, Arduino UNO board and related *Integrated Development Environment* (IDE), automation for software engineering design, at least of a source editor provided by a syntax highlighting and a cleaning debugging (e.g. NetBeans, Eclipse, Android Studio and Visual Studio) developed by an administrator of a console writing down Braille-like punched coding flowcharts often based on Unified Modelling Language translated by the complier switching plug-ins between various languages such as Java, C, C#, pearl, Phyton or sketched as visual programming in software for microscopy like OpenLab. Arduino Fundamentals exam demo (30 € https://create.arduino.cc/edu/courses/local/quiz/index.php) and example codes (https://www.arduino.cc/en/Tutorial/Blink).

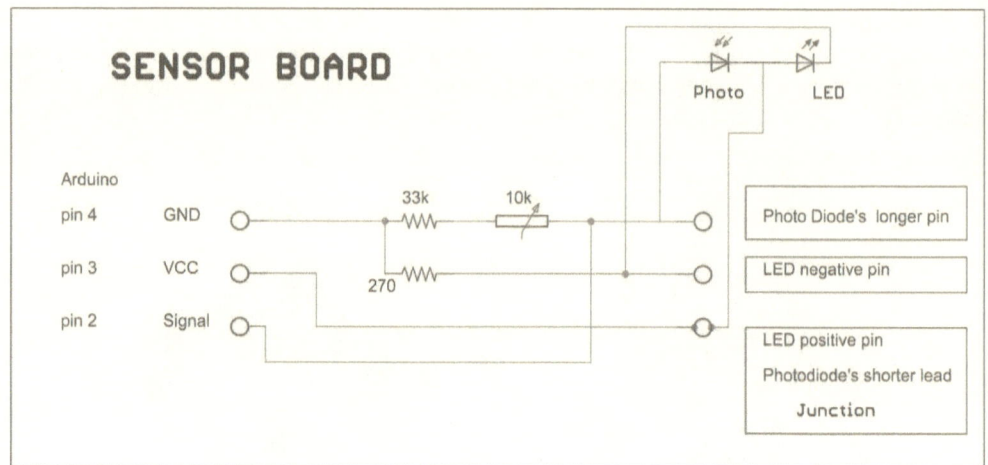

Figure 8 Optical Tachometer Board Schematic.

Among the main *Uranus* display technologies, the **Liquid Crystal Display** (LCD) produces images via light reflection and selective absorbance of bipolarized-filtered dying through anisotropic nano-beads or rods aromatic benzene rings pair, with polar and apolar ends, or an eutectic liquid solution (e.g. 4'-pentyl(heptyl, octyl, or alkyl oxy group) l-4-cyanobi (or tri)phenyl) in the background [e.g. *Cold Cathode Fluorescent Lamps* (CCFL), Width WLED, sub-pixels RGB-LED, Monochrome-LED] of a flat emissive panel (GEN 11: 2940 x 3370 mm), via its alignment due to the voltage application, for example, along helical pathways between two transparent electrodes (e.g. *Indium Tin-Oxide* InSnO) as in the *Twisted Nematic* (TN) in which can be placed a matrix of switching *Thin Films Transistors* (TFTs) [TFT/px], peculiar MOSFET having dielectric glass insulators substrates instead of semiconductors. In simpler words, the fluid motion obstacolate the light passage according the potential applied, filtering the radiation in each pixel manifesting the primary or combination of these colours according the TFT activated.

Less expensive, but both susceptible to ghost burns damages happening at T = 1200° C and highly energy demanding, **Plasma Display Panels** (PDP) (life-time t = 100 000 h = 4167 days = 11 years, power P = 500 W) works via a ionizing noble gaseous medium in which mercury (Hg) particles cause the quantum LOMO-HOMO excitation duality that involve the valence electrons of the phosphorous walls of two dielectric insulator plates with finger electrodes etched into them, for which the previous UV-light absorbance is followed by the releasing a lower energy ElectroMagnetic radiation in the range of PAR and even IR. The (super)-conductive stripes are parallel among them on the same plate, while, perpendicular with the other, making a screening network bordering the plasma state. The sparks flowing from one side to another switch on differentially the coloured sub-pixels forming the projected image onto the screen.

Field Emission Display (FED) based on the phosphorescent properties.

Advanced, but more expensive *Organic Light Emitting Diodes* (OLEDs) are based on electro- and photo-luminescence of biological molecules inversely proportional to the square of the size of dot according the inverse-square law. The following features specify LCDs: resolution expressed as number of column and rows (e.g. 1024 · 768), complete subset of gamut colour referred to the gamma correction based on the white point, contrast ration between bright light and dull dark.

Electro-emissive Quantum Dot Display (EQDP) uses luminescent nanocrystals semiconductors both inorganic (e.g. *Indium Selenide* InSe tuned in the PAR) and *Amorphous Matrix Organic Light Emitting Diodes* (AMOLED e.g. N,N'-Bis(3-methylphenyl)-N,N'-diphenylbenzidine) made via spinning *Centrifugal Rubber Mold Casting* (CRMC). QDP are about 75 times brighter than LC displays being characterized by a luminance $L_v = 4 \cdot 10^4$ nits (cd/m^2). Px = 0.26 mm.

Bistable multivibrator can used for quantum circuits in which superposition between two states is flip flopped in a continuous richer and lower parallel or poorer and faster series of shared information boundary interfaces in a chain in which the bit array stored is shifted through transitions overall bringing the name of *shift register*.

Transmission

Transmission (TX) is needed for connecting *'Zeus Genset'* to peripheral devices sending its powerful torque in a variable degree of rotation according the needed speed, direction using as occur in automatic systems, for example, fluid-dynamics valves of epicyclical planetary gearing or more modern real-time sensing microprocessor *Engine Control Units* (ECU) in which shift pattern and gear ratios are regulated by inductors. Smaller sprockets would be associated to faster speed at parity of torque. For e-motors multi-meter sensors for electrical parameters would be satisfying for setting a programmable ECU. Analogue or digital signals would modulate, check and correct the errors of the TX.

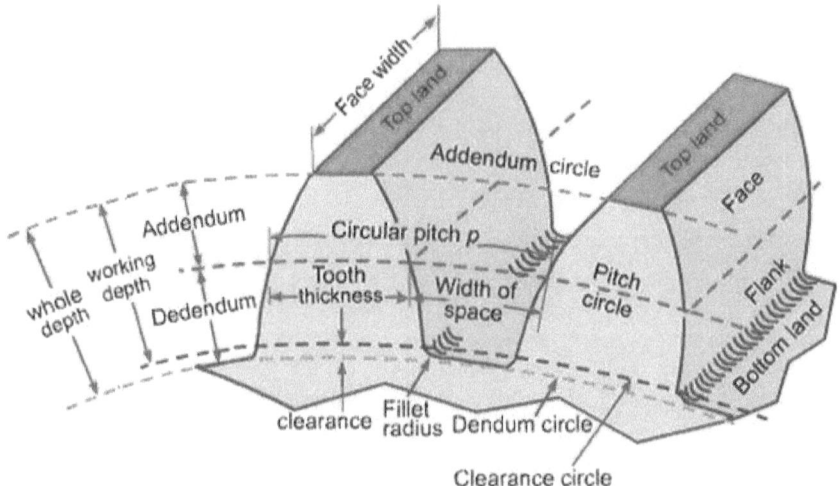

Figure 1 Gear nomenclature – Image Source : https://www.quora.com/

Gears diameter pitch :

$$\partial_{pitch} = \frac{N}{d} = \frac{N m_n}{\cos\varphi} = \frac{N \frac{p}{\pi}}{\cos\varphi}$$

Where:
N = number of teeth *Ares* 25, *Hera* 42, *Zeus* 48 (cogs)
α = helix angle = 30 °
m_n = module, length across the pitch diameter = 5.2 cm

p = circular pitch = 2 tooth thickness + width of the space between them
[m] d = standard pitch diameter [m]

Addendum:

$$a = \frac{D_o - D}{2}$$

Dedendum:

$$b = \frac{D - \bar{a}}{2}$$

Angular pitch :

$$1 < \theta_n = \text{---} = \text{---} < 360$$

Where:
N = number of teeth *Ares* 25, *Hera* 42, *Zeus* 48 (cogs)

Step angle = ¼ θ_n = 1.8 °

For proper fitting gears pair is to avoid gaps between the pitches, backlash mistake that could slow the motion or even change the direction of the spinning. The most modern and used velocity tooth profile is called **involute** of a circle, even if the cycloid is still utilized in some applications, because this evolvent curve of the roulette family (cycloids, epicycloids, hypocycloids, trochoids) if turned uniformly around its centre, then would lead the tangents to the involute having a given direction to move in a uniform translate uniformly. Precisely, this peculiar curve is done according the locus of points that differentiate according the mother function starting from one of its points (its exacerbation would be the involute of a line, otherwise, called perpendicular) depends only on the number of teeth on the gear, pressure angle, and pitch.

For a circle parametrically represented as (r cos(t); r sin(t)), the *arc length* (s) of an involute respect two properties:

$$c' s =$$
$$c(s) = k(s) \, n(s$$

Where:
c = curvature
n = normal

$$Path\ length = r\,(t - a)$$

Where:
a = segment of the mother function

Hence, the parametric equation of the *evolvent tangent* to the circle in the point P (x; y):

$$x\,t^{()} = r\,\cos t^{()} + {}^{(}t - a\,{}^{)}\sin t^{(())}$$
$$y\,t_{()} = r\,\sin t_{()} - t - a\,\cos t_{()}$$

That in *polar coordinate* would be:

$$x = r\,\cos\theta + \theta\sin\theta$$
$$y = r\,\sin\theta - \theta\cos\theta$$

For a = r · θ = 0 and 0 ≤ t ≤ t₂, the *arc length* is equal to:

$$s = t \cdot \frac{2}{3} \frac{r}{s}$$

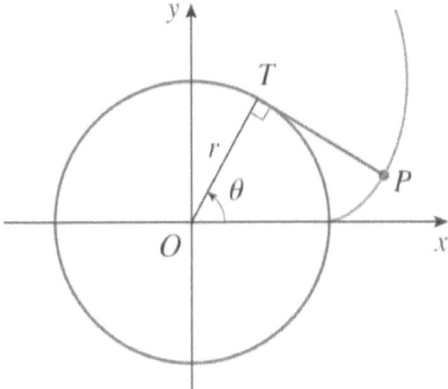

Figure 2 Involute of a circle - Image Source: https://www.youtube.com/watch?v=NIGw_dlEzQ4

The angular velocity ratio between two gears of a gearset must remain constant throughout the mesh. The teeth of pinon and wheel are tangent in a point that sliding along their opposite waves, describe the line of action. In other, words the tangential point position vary proportionally with the angle of rotation, for convention the pinon, in a typical range between the pitch points, better said segment of cogs contact. Decreasing the pressure angle, acute angle between the line of action and a normal to the line connecting the gear centres, commonly counting $\alpha = 20°$, provides lower backlash, smoother operation and less sensitivity to manufacturing errors. Hind legs of the plant hopper (*Issus coleopterous*) shows gear-like movement mechanisms.

24 : 1 L/D ratio, the high-carbon K100 steel *leadscrew* of adjustable RPM (single phase v = 50 Hz, P = 1.1 kW, AC, RPM = 1440 min⁻¹).

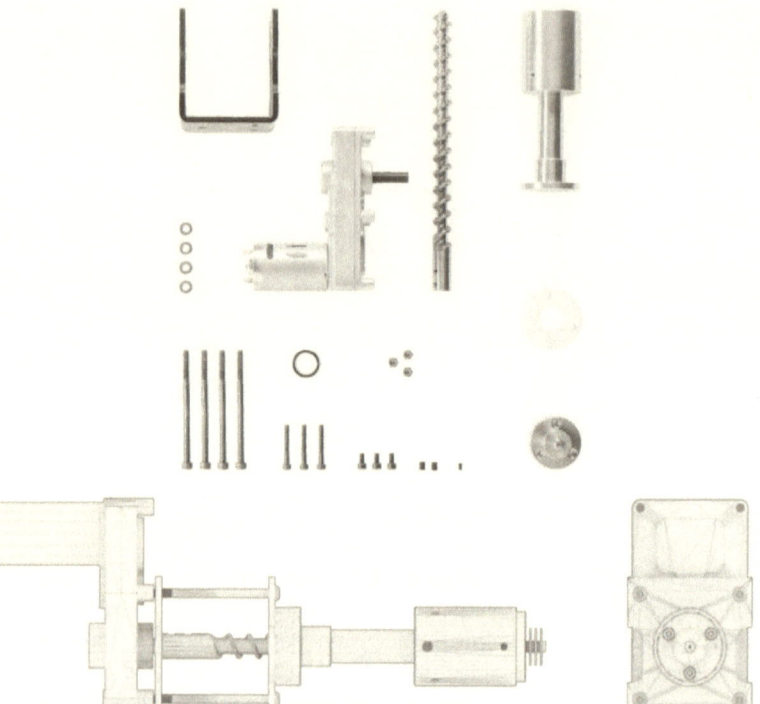

Figure *3* FelFil motor for insulated filament extruder with barrel (ø = 2 cm), 6 Chamfered flanges, toothed chain belt - Image Source: https://felfil.com/

'Olympus - the divine quadcopter' [(C) Copyright Antonio Silvestro, 2020] would be characterizedø by the presence of four *Zeus Gensets* øBLDC e-motor mode (20 < $m_{Olympus}$ < 140 g, can diameter = 8.5 mm and height h = 20 mm, shaft $_s$ = 1 mm and length l_s = 5 mm, JST type connectors, ΔV = 3

kV, 1.8 < I < 3.2 A, 4 motor mounts .stl, 35 < price < 80 €, $15 \cdot 10^3$ < RPM/ΔV < $17 \cdot 10^3$ Kv is inversely proportional with the length of the propellers (10 < l_p < 15 cm, m_p = 4 g, 4 € https://micro-motor-warehouse.com/products/cl-0820-15-11t).

Power [W]	Electricity [I]	Potential [ΔV]	Shaft diameter	Shaft length [m]	Cylinder diameter	Cylinder height [m]	Mass [m]	Euro [€]
5.8 < P < 9 kW	1.8<I<3.2A	ΔV = 3 kV		l_s = 5 mm		h = 20 mm	20 < $m_{Olympus}$ < 140 g	58

Antonio Silvestro right after having being involved in the Erasmus + project in Latvia as *'Solidary Corp'* volunteer helping children of the local community teaching Yoga, music, cooking and languages classes, doubtful to keep the MSc *'Plant, Food Sciences and Environmental Biotechnology'*, decided to divulge his holistic wisdom as *Kindle Self Publisher* (KDP) on Amazon, in the *'Santosha'* of holding a Ba degree in *"Biological Sciences"* got with honours and awareness of the need for tools only for going ahead in the R&D. Nevertheless, between the Bachelor and the Master, he travelled in Switzerland, UK and

Cyprus joining permaculture, agroforestry, aquaponics, and phytoremediation projects in countryside is developing communities, organic farms, designing environmental and wellness architectures, increasing his awareness through Yoga and Shamanism spiritual healing retreats, giving the *'HolYoga'* workshops in Denmark and Germany, both in private estates and in public festivals. Among the previous engineering designs the *'Telepika – Telescopic Pipe Moka'* of which the assembly protocol is available on Instructables and *'Aphrodite and Ares entwined in a torus inductor making the baby Hermes rototraslating!'* offered on Amazon KDP.